火 星

飞沙走石的星球

MARS

The Dusty Planet

（英国）埃伦·劳伦斯／著　刘 颖／译

江苏凤凰美术出版社

Copyright © 2014 Ruby Tuesday Books Ltd.

Chinese Simplified Character rights arranged through Media Solutions Ltd

Tokyo Japan email: info@mediasolutions. jp, jointly with the Co-Agent of

Gending Rights Agency (http://gending. online/).

Simplified Chinese edition copyright

©2025 JIANGSU PHOENIX FINE ARTS PUBLISHING, LTD

All rights reserved.

著作权合同登记图字：10-2022-144

图书在版编目（CIP）数据

火星：飞沙走石的星球 /（英）埃伦·劳伦斯著；

刘颖译 . -- 南京：江苏凤凰美术出版社，2025. 1.

（环游太空）. -- ISBN 978-7-5741-2027-3

Ⅰ . P185.3-49

中国国家版本馆 CIP 数据核字第 2024PQ1549 号

策　　　划	朱　婧	
责 任 编 辑	高　静　奚　鑫	
责 任 校 对	王　璇	
责任设计编辑	樊旭颖	
责 任 监 印	生　嫄	
英 文 朗 读	C.A.Scully	
项 目 协 助	邵楚楚　乔一文雯	

丛 　书 　名	环游太空
书 　　　名	火星：飞沙走石的星球
著 　　　者	（英国）埃伦·劳伦斯
译 　　　者	刘　颖
出 版 发 行	江苏凤凰美术出版社（南京市湖南路1号 邮编：210009）
印 　　刷	南京新世纪联盟印务有限公司
开 　　本	710 mm×1000 mm　1/16
总 印 张	18
版 　　次	2025 年 1 月第 1 版
印 　　次	2025 年 1 月第 1 次印刷
标 准 书 号	ISBN 978-7-5741-2027-3
总 定 价	198.00 元（全 12 册）

版权所有　侵权必究

营销部电话：025-68155675　营销部地址：南京市湖南路 1 号
江苏凤凰美术出版社图书凡印装错误可向承印厂调换

目录  Contents

书中加粗的词语见词汇表解释。

Words shown in **bold** in the text are explained in the glossary.

欢迎来到火星
Welcome to Mars

2021年2月18日*，"毅力号"火星车成功在火星着陆。

On February 18, 2021, a robotic rover called *Perseverance* landed on Mars.

为了抵达火星，它在太空中旅行了将近4.83亿千米。

To reach Mars, it had flown almost 483 million kilometers.

着陆后，"毅力号"很快着手拍摄了超过75 000张照片并将其传回地球。

The rover soon got busy taking more than 75,000 photos and sending them back to Earth.

它也录下了火星上风吹过的声音和自己的轮子在火星那布满岩石和尘土的地面上行驶的声音。

It also recorded the sounds of Martian winds blowing and its own wheels rolling over the rocky, dusty land.

随后于2021年6月1日，"毅力号"开始了自己的第一项科学勘测任务。

Then on June 1, 2021, *Perseverance* began its first science mission.

约1 090万人参加了"把你的名字送上火星"活动。三块承载着这些人名字的微型芯片被装载在"毅力号"上。

About 10.9 million people joined a "Send Your Name to Mars" project. Their names are on three tiny computer chips onboard *Perseverance*.

*译者注：本书时间为美东时间。

在火星上，寒风吹过。这风比地球北极12月的风还要冷。
On Mars, icy winds blow and it's colder than the North Pole in December.

"毅力号" *Perseverance*

火星地表 The surface of Mars

火星土壤由于富含铁锈，所以看上去是红棕色的。
The **soil** on Mars looks reddish-brown because it contains lots of **rust**.

太阳系 The Solar System

火星以近87 000千米每小时的速度在太空移动。

Mars is moving through space at nearly 87,000 kilometers per hour.

它围绕着太阳做一个巨大的圆周运动。

It is moving in a big circle around the Sun.

火星是围绕太阳公转的八大行星之一。

Mars is one of eight **planets** circling the Sun.

八大行星分别是水星、金星、我们的母星地球、火星、木星、土星、天王星和海王星。

The planets are Mercury, Venus, our home planet Earth, Mars, Jupiter, Saturn, Uranus, and Neptune.

冰冻的彗星和被称为"小行星"的大块岩石也围绕着太阳公转。

Icy **comets** and large rocks, called **asteroids**, are also moving around the Sun.

太阳、行星和其他天体共同组成了"太阳系"。

Together, the Sun, the planets, and other space objects are called the **solar system**.

太阳系中的大多数小行星都集中在被称为"小行星带"的环状带中。

Most of the asteroids in the solar system are in a ring called the asteroid belt.

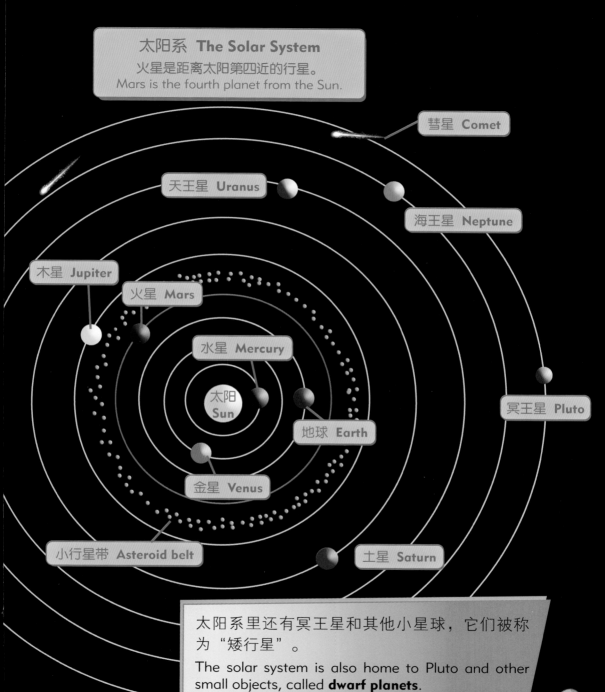

太阳系 **The Solar System**
火星是距离太阳第四近的行星。
Mars is the fourth planet from the Sun.

彗星 **Comet**

天王星 **Uranus**

海王星 **Neptune**

木星 **Jupiter**

火星 **Mars**

水星 **Mercury**

太阳 **Sun**

地球 **Earth**

冥王星 **Pluto**

金星 **Venus**

小行星带 **Asteroid belt**

土星 **Saturn**

太阳系里还有冥王星和其他小星球，它们被称为"矮行星"。

The solar system is also home to Pluto and other small objects, called **dwarf planets**.

7

火星的奇幻之旅
Mars' Amazing Journey

行星围绕太阳公转一圈所需的时间被称为"一年"。

The time it takes a planet to **orbit**, or circle, the Sun once is called its year.

地球绕太阳公转一圈需要略多于365天，所以地球上的一年有365天。

Earth takes just over 365 days to orbit the Sun, so a year on Earth lasts 365 days.

火星比地球离太阳远得多，所以它绕太阳公转的路程更长。

Mars is farther from the Sun than Earth, so it must make a much longer journey.

火星绕太阳公转一圈大约需要687个地球天。

It takes Mars about 687 Earth days to orbit the Sun.

这意味着，火星上的一年几乎是地球一年的2倍！

This means that a year on Mars is nearly twice as long as one on Earth!

火星 Mars

当行星围绕太阳公转时，它也像陀螺一样自转着。

As a planet orbits the Sun, it also spins, or **rotates**, like a top.

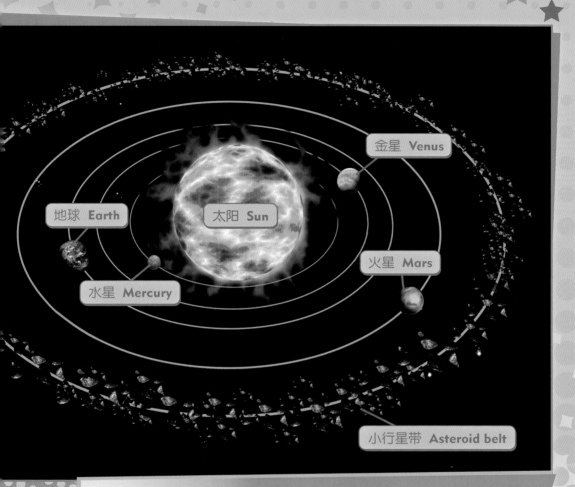

地球绕太阳公转一圈的路程约为9.4亿千米，而火星绕太阳公转一圈的距离约为14亿千米。

To orbit the Sun once, Earth makes a journey of about 940 million kilometers. Mars must make a journey of about 1.4 billion kilometers.

近距离观察火星
A Closer Look at Mars

火星是地球在太空里最近的邻居之一，但它与地球截然不同。

Mars is one of Earth's closest neighbors in space, but it is a very different world.

火星表面看起来像是一片飞沙走石的沙漠。

The surface of Mars looks like a dusty, rocky **desert**.

火星也是一颗风暴猛烈的星球。

Mars is also a very stormy planet.

火星上的疾风能将沙尘卷上天空。

The planet's fast winds blow dust up into the sky.

这让火星的天空呈现出橘粉色。

This makes the sky look pinkish-orange.

有时，整颗行星都被一场巨大的沙尘暴所笼罩。

Sometimes, the whole planet is covered with one giant dust storm!

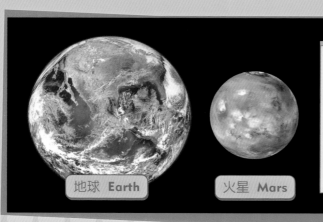

地球 Earth

火星 Mars

火星比地球小。如果将地球的体积比作棒球的话，那么火星就相当于乒乓球。

Mars is smaller than Earth. If Earth were the size of a baseball, Mars would be the size of a ping pong ball.

火星上的这场沙尘暴是由一个在火星上空飞行的航天器拍摄的。
This dust storm on Mars was photographed by a spacecraft flying above the planet.

这张照片显示了火星上一场巨大的沙尘暴。沙尘暴即充满尘埃的旋风。
This picture shows a huge dust devil on Mars. A dust devil is a whirlwind made of dust.

太阳系的 "纪录创造者"
Solar System Record Breakers

火星有太阳系所有行星里最高的山峰和最深的峡谷。

Mars is home to the tallest mountain and the deepest **canyon** on any planet in the solar system.

这座山是一座高约22千米的巨型火山。

The mountain is a giant **volcano** that is about 22 kilometers high.

这座火山太大了，如果放在地球上，它可以覆盖整个美国亚利桑那州。

It's so large that if it were set down on Earth, it would cover all of Arizona.

火星上还有一条宽达200千米的巨大峡谷。

Mars' gigantic canyon is 200 kilometers wide.

它的深度是美国大峡谷的3倍！

It is three times as deep as the Grand Canyon!

水手峡谷
Valles Marineris canyon

火星上的大峡谷被称为水手峡谷。

Mars' monster canyon is called Valles Marineris (VAL-iss mare-ih-NARE-us).

火星上的大火山被称为奥林匹斯山。

Mars' massive volcano is called Olympus Mons (oh-LIM-pus MAHNS).

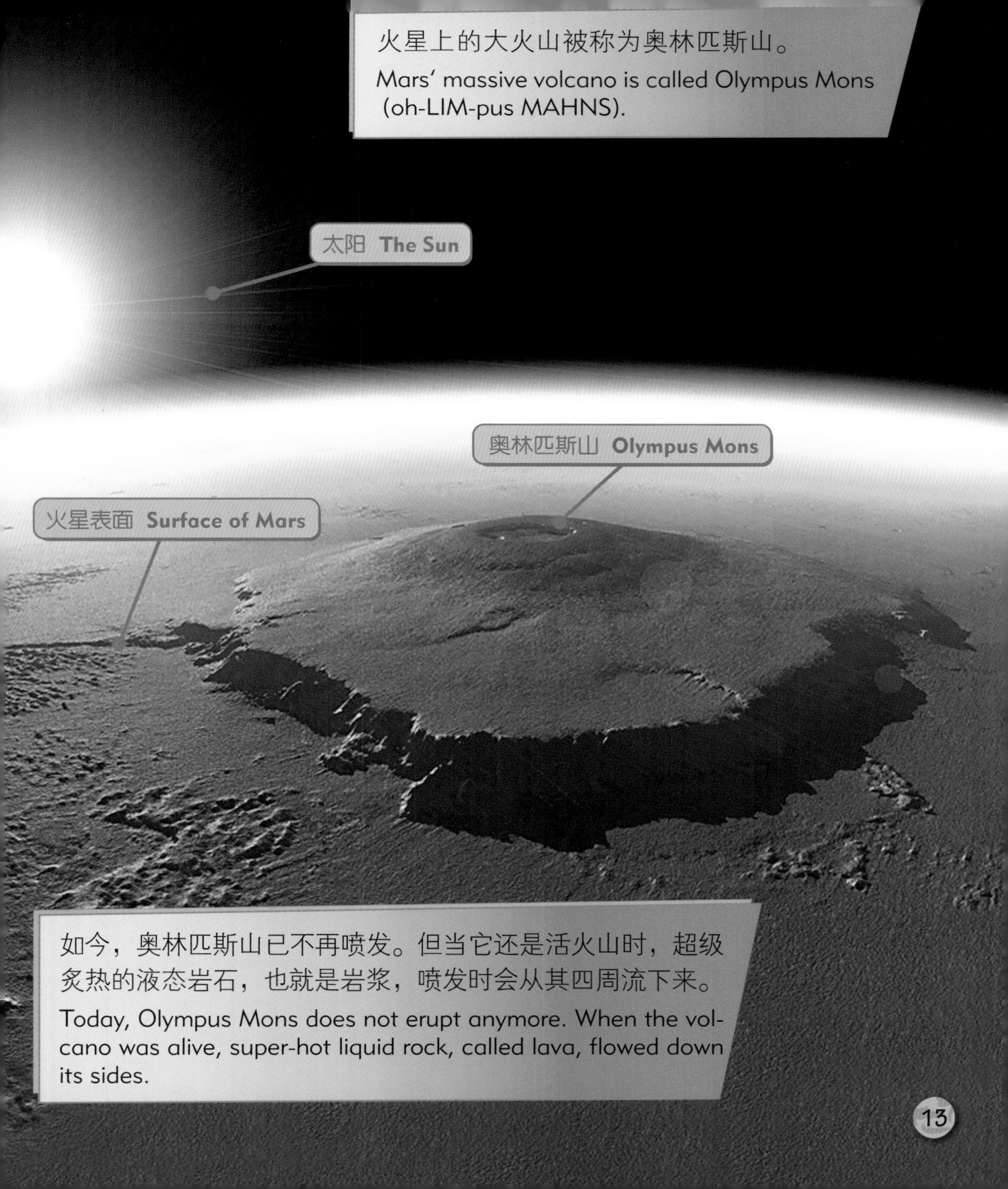

太阳 **The Sun**

奥林匹斯山 *Olympus Mons*

火星表面 **Surface of Mars**

如今，奥林匹斯山已不再喷发。但当它还是活火山时，超级炙热的液态岩石，也就是岩浆，喷发时会从其四周流下来。

Today, Olympus Mons does not erupt anymore. When the volcano was alive, super-hot liquid rock, called lava, flowed down its sides.

火星的 "小伙伴们"
Mars' Tiny Companions

火星有两个 "小伙伴"。在火星围绕太阳公转时，它们一直 "形影不离"。

Mars has two tiny companions that stay close to the planet as it orbits the Sun.

这两颗岩质卫星分别是福波斯（火卫一）和得摩斯（火卫二）。

These two rocky moons are called Phobos (FOH-bohss) and Deimos (DEE-mohss).

不同于地球的球形卫星，火星的卫星像土豆块。

Unlike Earth's Moon, which is round, Mars' moons look like lumpy potatoes.

福波斯和得摩斯围绕火星公转。

Phobos and Deimos are orbiting Mars.

福波斯距离火星很近，它公转一圈仅需8小时。

Phobos is so close to Mars, it takes just eight hours to orbit the planet.

得摩斯距离火星较远，它公转一圈需要30小时。

Deimos is farther away and orbits Mars once every 30 hours.

得摩斯的长度只有约14千米。
Deimos is only about 14 kilometers across.

火星最大的卫星——福波斯，长度只有约27千米。
Mars' largest moon, Phobos, is only about 27 kilometers across.

探测火星的任务
Missions to Mars

科学家已发射数个航天器和探测器去研究火星。

2012年8月，一个约有汽车大小的火星探测器"好奇号"在火星着陆。

该探测器在火星上的岩石内部发现了能证明火星上曾经有水源存在的证据。

生物需要水才能存活。

这项发现意味着火星可能存在过微生物！

Scientists have sent spacecrafts and robots to Mars to study the planet.

In August 2012, a car-sized robot called *Curiosity* landed on Mars.

The rover has found clues in rock that show there was once water on the planet.

Living things need water to survive.

This means that tiny living things called **microbes** may once have lived on Mars!

"好奇号"装有摄像头、挖掘工具以及其他科学设备。它能用钻头钻入岩石，并把钻出的岩石粉末收集起来化验。

Curiosity has cameras, tools, and science equipment on board. The robot drills into rocks. Then it collects and tests the powdered rock it has made.

岩石粉末 Powdered rock

16

火星上并没有海洋或是河流。但在2018年，名叫"火星快车号"的空间探测器发现了在火星南极的冰层下可能有湖泊存在。

Mars doesn't have oceans or rivers. But in 2018, a spacecraft named *Mars Express* detected signs that there could be lakes of water beneath the ice at Mars' south pole.

"好奇号" *Curiosity*

地球上的科学家使用电脑远程指挥"好奇号"移动和工作。科学家们的命令需要大约21分钟才能从地球传到火星上的"好奇号"。

Scientists on Earth use computers to tell *Curiosity* where to go and what to do. It can take up to 21 minutes for the scientists' commands to reach the rover from Earth.

"好奇号"会给自己拍摄自拍并传给地球上的科学家，这样他们就知道"好奇号"当前是否运转良好，有没有损坏！

Curiosity sends selfies back to Earth so scientists can be sure it is in good shape!

"毅力号"与"机智号"
Perseverance and Ingenuity

和"好奇号"一样，"毅力号"火星车也在寻找很久以前存在过的火星微生物的痕迹。

Like *Curiosity*, the robot rover *Perseverance* is searching for signs of Martian microbes from long ago.

"毅力号"也会收集火星上的岩石和土壤样本，并把它们都储存在特制软管里。

Perseverance will also gather rock and soil samples and store them in special tubes.

在将来的某一天，另一项前往火星的任务会收集这些储存的样本并将其送回地球。

One day, a future mission to Mars may collect the samples and return them to Earth.

"毅力号"携带了一架小型太阳能直升机——"机智号"，前往火星。

Perseverance carried a tiny solar-powered helicopter, called *Ingenuity*, to Mars.

它是人类第一架在别的星球上空成功飞行的飞行器。

It's the first craft to ever fly around on another planet!

"机智号"重约1.8千克。它被安置在"毅力号"的腹部，随其一起到了火星。这张图片显示了火星车把"机智号"放在火星地表的场景。

Ingenuity weighs just 1.8 kg. It flew to Mars in *Perseverance*'s belly. Here, the rover lowers *Ingenuity* onto the surface of Mars.

"机智号" *Ingenuity*

总有一天，像"机智号"这样的直升机将会被用来探索那些对于火星车来说太陡太滑的地方。

One day, helicopters like *Ingenuity* could be used to explore places that are too steep or slippery for rovers.

飞行的"机智号"
Ingenuity in flight

"毅力号" *Perseverance*

如果人类真的有一天要前往火星，那他们需要氧气来呼吸。"毅力号"正在测试一个叫作"MOXIE"的工具——它能将火星上的二氧化碳气体转化成氧气。

If humans one day visit Mars, they will need **oxygen** to breathe. *Perseverance* is testing a tool called MOXIE. This tool can turn **carbon dioxide** gas on Mars into oxygen.

有趣的火星知识
Mars Fact File

以下是一些有趣的火星知识：火星是距离太阳第四近的行星。

Here are some key facts about Mars, the fourth planet from the Sun.

火星的发现
Discovery of Mars

不用望远镜也能在天空中看见火星。人们早在古代就发现了火星。

Mars can be seen in the sky without a telescope. People have known it was there since ancient times.

火星是如何得名的
How Mars got its name

火星是以古罗马战神的名字命名的。

Mars is named after the Roman god of war.

行星的大小
Planet sizes

这张图显示了太阳系八大行星的大小对比。

This picture shows the sizes of the solar system's planets compared to each other.

水星 Mercury
地球 Earth
天王星 Uranus
太阳 Sun
木星 Jupiter
金星 Venus
火星 Mars
土星 Saturn
海王星 Neptune

火星的大小
Mars'size

火星的直径约6 779千米

6,779 km across

火星自转一圈需要多长时间
How long it takes for Mars to rotate once

约25小时

Nearly 25 hours

火星与太阳的距离
Mars'distance from the Sun

火星与太阳的最短距离是206 655 215千米。

火星与太阳的最远距离是249 232 432千米。

The closest Mars gets to the Sun is 206,655,215 km.

The farthest Mars gets from the Sun is 249,232,432 km.

火星围绕太阳公转的平均速度
Average speed at which Mars orbits the Sun

每小时86 677千米

86,677 km/h

火星绕太阳轨道的长度
Length of Mars' orbit around the Sun

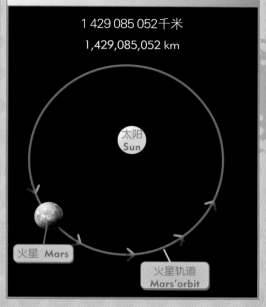

1 429 085 052千米

1,429,085,052 km

太阳 Sun

火星 Mars

火星轨道 Mars'orbit

火星的卫星
Mars'moons

火星有两颗卫星
Mars has two moons:

福波斯 Phobos

得摩斯 Deimos

火星上的一年
Length of a year on Mars

将近687个地球天

687 Earth days

 ## 火星上的温度
Temperature on Mars

最高温度：超过20摄氏度

最低温度：低于零下150摄氏度

Highest: over 20°C

Lowest: below 150°C

动动手吧：制作探测机器人
Get Crafty : Build a Robot Explorer

快动手制作属于自己的探测机器人吧！

机器人可能由以下物品组成：
- 空盒子、纸箱或瓶子
- 卷筒纸的纸筒
- 使用过的干净铝箔
- 你能想到的其他任何材料

这是一台漫游车的参照模型，但你可以自行设计。设计前请思考以下问题：
- 你设计的漫游车将会是什么样子？
- 它将如何移动？
- 它有哪些工具和设备？
- 它可能执行什么任务？

你需要：
- 剪刀（在成年人的帮助下使用）
- 胶水
- 胶带
- 油漆和画笔

词汇表 Glossary

小行星 | asteroid
围绕太阳公转的大块岩石，有些小得像辆汽车，有些大得像座山。

峡谷 | canyon
陡峭的深谷。

二氧化碳 | carbon dioxide
地球上一种无形的气体，人和动物都会呼出。

彗星 | comet
由冰、岩石和尘埃组成的天体，围绕太阳公转。

沙漠 | desert
布满岩石或沙砾的地方，无水或缺水。有些沙漠很炎热，另一些却很寒冷。

矮行星 | dwarf planet
围绕太阳运行的圆形或近圆形天体，比八大行星小得多。

微生物 | microbe
极小的生物，无法用裸眼看见。能够使人生病的细菌就是一种微生物。

公转 | orbit

围绕另一个天体运行。

氧气 | oxygen

空气中一种无形的气体，是人类和其他动物呼吸所必需的。

行星 | planet

围绕太阳公转的大型天体：一些行星，如地球，主要是由岩石组成的；其他的行星，如木星，主要是由气体和液体组成的。

自转 | rotate

物体自行旋转的运动。

铁锈 | rust

一种在金属表面形成的橘红色疏松物质。

土壤 | soil

覆盖地面的疏松物质，存在于许多地方。主要由细小的岩石构成。

太阳系 | solar system

太阳和围绕太阳公转的所有天体，如行星及其卫星、小行星和彗星。

火山 | volcano

地下岩浆喷出地表形成的山丘，部分火山会有高温的液态岩石和气体从开口处喷发；存在于行星或其他天体上。